The BRIDGE

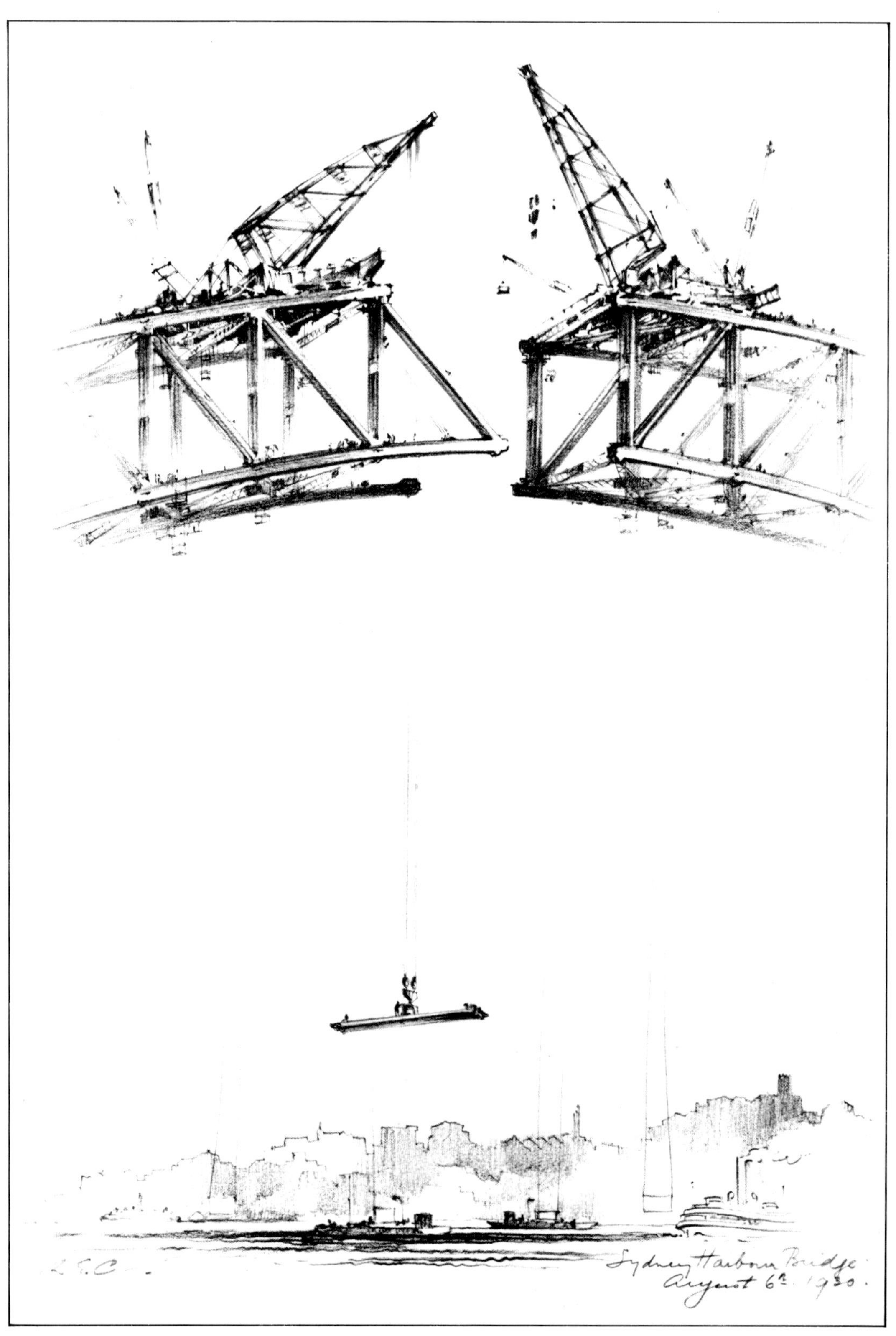

Sydney Harbour Bridge.
August 6th 1930.

The BRIDGE

Robert Emerson Curtis

The Currawong Press

Published by The Currawong Press Pty Ltd
PO Box 233, Milson's Point, NSW 2061

First published 1981
c Copyright R.E. Curtis 1981
Produced in Australia
Typeset by Familiar Faces Typesetting
77 Willoughby Rd, Crows Nest, 2065
Printed in Hong Kong

ISBN 0 908001 25 8

Foreword

Those of us who witnessed the building of the Sydney Harbour Bridge were
caught up unconsciously in the excitement of that enormous project. For that
generation, the Bridge's outline was inescapable. It dominated the city and it
altered Sydney's character.

When the Bridge was finally opened — and what a dramatic episode that was! —
some people had secret misgivings that the whole structure was on too grand a
scale, that the expanse of Bradfield Highway set apart for motor traffic was extra-
vagantly excessive and would never be needed. It may have looked that way during
the Depression years of the early 1930s, but it certainly doesn't fifty years later,
when public clamour for another Harbour crossing is mounting every year.

I am delighted that a new generation of Sydneysiders will now be able to enjoy
these distinguished drawings by Robert Emerson Curtis. He was a privileged pictorial
recorder of the Bridge's creation. Through his artistry, our children and grand-
children may experience the excitement that we felt during those years before 1932.

And for those who watched this huge construction changing Sydney's skyline,
Mr Curtis's book is certain to re-awaken a flood of treasured memories.

Philip Geeves OAM
Fellow, Royal Australian
Historical Society.

Preface

January, 1928 — A gust of summer rain was sweeping across the 'Heads' as we entered Port Jackson. 'A summer shower,' I said, hoping to cheer my wife and small daughter, for this was their first sight of Sydney Harbour of which they had heard so much. Just as suddenly as it had begun, the rain stopped and the sun emerged to brighten green and rocky headlands, sandy beaches and high garden slopes dotted with cool and elegant homes. There were sweeping impressions of red-tile roofs, of anchored ships and idle yachts. Then, as if floating towards us, the island rock of old Fort Denison appeared, just as a Manly-bound ferry crowded with waving school-children cruised past. I had longed to see such sights again after years overseas, but I knew that Sydney's summer climate could not guarantee rainless skies. Even as the distant city came into focus, darkening clouds were gathering. There was just time for a fleeting glimpse of Circular Quay — all bustle with ferries and trams — before a grey curtain dropped as the rain pelted, down. Very slowly we rounded Dawes Point, nothing visible as we edged into Darling Harbour to moor.

As a genial Customs officer checked our baggage, he mentioned that Sydney was getting on with the bridge. 'What Bridge?' we asked. He seemed stunned by our ignorance. Hadn't we heard that Sydney was building the greatest arch-bridge in the world? Not much to be seen yet, he added, but they were well ahead with the approaches. It was stirring news — a harbour bridge at last. The more I thought about it, the more I wanted to get involved with the story of its construction, to record a grand and important event. In the following years many opportunities were provided for me to work on the site, both in the workshops and on the arch itself. In due course some of my lithographs were published.

In March of 1932 — after eight years of building — the Bridge was completed. But the opening celebrations took place against a grim backdrop. A terrible world

Depression was at its worst, with countless hundreds of thousands of men and women out of work and Australia was no exception. If there was excitement and triumph in the completed Bridge hopelessness was ever present in the wings. Not all approved. There were those who angrily counted the cost. Others were critical of the design and some even despaired for the very future of Sydney. 'The Bridge is a tragedy . . . it is 50 years ahead of its time . . . serves no necessary purpose and reflects the lack of stability of the Sydney people,' was the comment of one distraught businessman.

Now, half a century has passed. From notes and drawings made during those years I worked on the Bridge, and with many drawings not previously published, I again have a chance to tell something of this great achievement.

THE BRIDGE — really is a story of STEEL, GRANITE and GUTS and I hope it captures something of the jubilation and humour as well as the thought, courage and muscle of those who built the bridge.

Robert Emerson Curtis
Sydney

The Bridge

What is it about building a bridge that so captures the imagination? Is it the logic of its design? Its apparent lightness, even fragility? Its exposed anatomy as we watch its construction? Or is it the boldness of such a project and the inherent risks that must be taken? And can the building of a bridge be accepted as a work of Art? One hears today that there can be little creative inspiration in the subject of *work,* since machines, computers and their micro-chips can now do so much that they have transcended the labours of the worker and the craftsman — those men and women who wield the tools. But such reasoning takes little account of the great *work* achieved by labour today in cities, dams, bridges, giant towers and oil-rigs, the ships, docks and aircraft et al. However clever the computer may be, as long as man has a brain and the muscle to respond to it, the possibilities of *work* as a theme for Art remain unlimited.

At the time of the construction of the great Sydney Harbour Bridge it was undoubtedly an inspiration to thousands of Australians. The work was visible at every stage as the giant arms stretched slowly towards each other across the water: the weaving of the lace-like panels of steel against the sky; the hoisting aloft of every beam and girder of the bridge from water-level, perhaps a rigger or two perched casually at an end. For those of us who witnessed such sights, whether in early morning light, the full glare of the sun, in high wind or in rain, in a glowing dusk or a foggy morning, they were to stamp images in our memory that could never be erased.

Bridge building, of course, is as old as time. A fallen tree across a stream would have prompted primitive man to realise there was no need to wait until a tree collapsed of its own accord. Dwellers of rain forests and jungles built crude rope-bridges of twisted cane and vine which was to lead to the principle of the suspension bridge. The early log bridge developed into the cantilever. Logs firmly anchored on opposite banks of a stream with their ends projecting over the water provided a platform upon which to lay a centre log to span the intervening space, a primitive cantilever. The Romans are believed to have been the first arch builders. They invented, or adopted the principle of the arch because it made stone bridges possible, strong enough to support a roadway. The "Senator's Bridge," built in Rome in 181 B.C., lasted until 1598 before two of its arches collapsed. Yet one of the arches still survives today.

The decision to build a steel suspension bridge over the Niagara River between the United States and Canada was made as far back as 1840. A reward of five dollars was offered to anyone who could get a line across the gorge. Hundreds of boys tried with their kites. On the first windy day, a boy claimed the reward by flying his kite to the opposite bank. That first light line was used to eventually draw a rope across the gorge. It would have taken more than a kite to span the Golden Gate

into San Francisco Harbour which claims the longest of all single-span suspension bridges in the world. The most notable crossing to be bridged with the cantilever principle was Scotland's great Firth of Forth Railway bridge in 1890. It is more than 2.4 kilometres long.

When the Harbour Bridge was completed in 1932, Sydney had built the greatest and the widest single-arch Bridge in the world. She had waited a long time for her first major Harbour crossing but by the time it was finished it made engineering history, for the great arch had been built out from both shores without a single support from the water below during its construction. How this was done at a time when many engineers believed it would be impossible is described later.

Some of us still recall commuting to Sydney before the great bridge was opened—before today's packed trains, crowded buses, and the morning and evening 'peak-hour crawl' of impatient motorists. It was a relaxing half-hour for all as we were ferried restfully, morning paper at hand, across the fresh and sparkling waters of

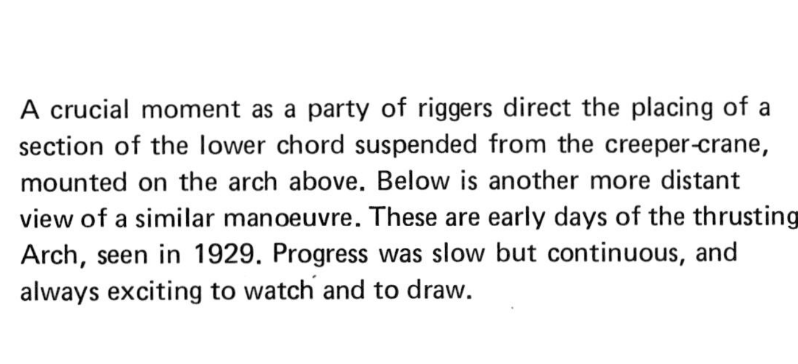

A crucial moment as a party of riggers direct the placing of a
section of the lower chord suspended from the creeper-crane,
mounted on the arch above. Below is another more distant
view of a similar manoeuvre. These are early days of the thrusting
Arch, seen in 1929. Progress was slow but continuous, and
always exciting to watch and to draw.

THEY'RE BUILDING A BRIDGE

Those men who live on our street
Sitting on the steps
Smoking and yarning in the dusk
Those men are building a bridge.

Those men are building a bridge,
Slowly it grows — slowly
Sweat and steel; steel and fire
Reaching out across the water.

Beams swinging from a crane
Hands tugging; muscles straining
Voices bellowing
Hammers pounding red-hot rivets
Burning into steel.

A man slips — is caught — curses —
Little ferries far below
Rip to shreds a shadow-web of steel
Eyes stare anxiously — mouths open
Watching those midget men high aloft.

Curse the wind, the cold, the heat
Curse and growl. Strike!
Strike for shorter shifts;
Strike for danger-money.
Who cares?
Let the mob use the ferries!
A day or two — or maybe three
And the bridge lies quiet
Under the sun — waiting.

The men come back
Losing a quid or making a quid
'Wot the hell!'

The bridge goes on
And as it grows
Something of each man is bolted to it
Steel unto steel
Something of himself he can never take back
Born of his toil — his guts — his pride
The bridge is part of him
His work — a curse — and a glory!

At dusk
They smoke and yarn and argue
Sitting on the steps down our street
Those men who are building a bridge.

The cables which hold the cantilevered "half-arch" in position pass underground, encircling Milson's Point and are attached again at the top of the corresponding end-post of the Bridge

Bands of steel cables were laid underground through horse-shoe shaped tunnels on both sides of the harbour. These were attached to the end posts of the arch support for each half out over the water until they were united in the centre. This completed the first stage of building the bridge.

The quaint old timber and iron ferry wharves
that lined the Quay were not replaced and
updated until many years after the Bridge
was built.

the harbour. Perhaps the sight of a single ferry trailing its silver wake through the
dark shadow of the Bridge will prod the memory of a time when the harbour was
astir with fleets of ferry-boats, large and small. When the morning voyage from
Manly by the ocean or from sheltered Watson's Bay or a run down the Lane Cove
or Parramatta rivers had almost the carefree atmosphere of a holiday. Part of the
fun was scampering down the hill for the boat at Mosman or Cremorne, at Kirri-
billi or Milson's Point, for that final stretch across the harbour to scramble ashore
down a clattering ramp at Circular Quay.

The old horse-punts have disappeared forever, those vehicular ferries that criss-
crossed the main shipping lane, protesting at priviledged liners and attendant tugs,
tankers, trawlers, scows and even private yachts. Such is our bond with the auto-
mobile and the fear of a wasted minute that comparatively few ferries still work
the harbour now — together with a trio of wave-skipping hydrofoils that make
the sedate but comfortable ferry ride to Manly seem like a voyage into history.

The main business of the harbour today is no longer for the ferry-boat commuter
but the monstrous container ships, oil tankers and tugs, for police and fire launches,
Navy repair boats and submarines. Though fewer great cruise-ships now exist, they
are not infrequent visitors. However, it is at the week-end that the waters really
come alive, dancing to the surge and sweep of hundreds of fully rigged racing
yachts, their spinakers a swirling riot of colour as they turn downwind.

Long before 1932 several small bridges over deep-water inlets of the upper
harbour provided the shortest road link between the heart of the city and its

Workday or holiday there was always a feeling
of freedom from care around the Quay. Travel
habits and life-styles have changed since this
sketch was first reproduced — when electric
trams and ferry-boats were Sydney's most
popular transport. To many, the Quay still has
that relaxed feeling. Perhaps every great and
fast growing city should have such a corner
with a magnificent view and a place to relax.

northern shore, a distance of 20 kilometres. It was known as the Five Bridges Road, which crossed inlets at Pyrmont, Glebe, Iron Cove, Gladesville and finally Fig Tree on the Lane Cove river. For a century and a half the main body of the harbour resisted all schemes and every carefully drawn plan to bridge it. The simplest way was to cross by ferry and, long before the coming of steam-power and the first paddle-wheeler, crossings were conducted by the Government-licensed watermen. William Blue, better known as "Billy Blue", is the most affectionately remembered. He was a part negro of Jamaican origin who, after having served his term as a convict, had been granted land and a ferryman's license by Governor Macquarie. He was able to announce in the *Sydney Gazette* that he could offer: 'a tight and clean boat, an active oar and an unalterable inclination to serve all who honoured him with their commands!' He and the sons who followed him are remembered today by the northern headland named Blue's Point. James Milson, an enterprising free immigrant, took up a grant of 20 hectares on the northern bank of the harbour and started a farm, but found more profit in quarrying sandstone and punting it across the water to sell it for ship's ballast. He too was granted a waterman's license and started a ferry service from Milson's Point to the eastern side of Sydney Cove.

In 1817, a somewhat fanciful scheme to bridge the gap between Dawes Point and Milson's Point came to the notice of Governor Macquarie, the first ardent and determined developer of the Colony. The scheme's author was Francis Greenway, a civil architect, exiled from England to serve a term of 14 years in the Colony for forging part of a building contract. His bridge proposal fell on deaf ears, but Macquarie recognised Greenway's ability and soon put him to work on more urgently needed tasks. Appointed a government architect at three shillings (30 cents) a day his earliest achievement was the design of Australia's first lighthouse on Sydney's southern headland, a structure so aesthetically pleasing that it gained the convict architect a pardon.

The morning tonic! Not a line of the daily news
is missed as a harbour ferry speeds us across the
sparkling water!

I felt the curious eyes of the inhabitants of North Sydney's Arthur Lane upon me as I struggled to complete this drawing of the Dawes Point pylon. A passenger ferry below is leaving Lavender Bay as one of the old vehicular punts pulls out from the southern shore.

A total of 802 buildings were demolished to clear the way for the rail and road approaches to the Bridge. Most houses on the city side were located in the notorious Rocks area and had been resumed by the State during an outbreak of bubonic plague in 1901. The old Scots Church, erected in 1826 by Dr. Dunmore Lang, was pulled down. On the northern side of the harbour, 469 buildings disappeared.

The early-morning stampede at Milson's Point for a city-bound ferry.

Preserved in Sydney's archives are a number of early Bridge plans that never left the drawing board. We should be grateful on both counts! The first was a low wooden structure on piled foundations that would have completely sealed off the upper harbour to shipping. Later came a 'swing' bridge, followed by a seven-span truss bridge which in turn was followed by several plans for cantilever and suspension bridges as well as subways and tunnels. The most extraordinary proposal of all, however, was to cut a grand shipping canal from the head of Neutral Bay on the north side to nearby Lavender Bay, bridging the canal across the high land of North Sydney and using the excavated rock to build a low roadway wall over the main harbour between Dawes Point and Milson's Point!

Though a great deal of time and thought and money was spent in planning what was yearly becoming a more urgent need, it was not until the early years of the present century that a young Queensland-born engineer by the name of John Job Crew Bradfield entered the field that things began to stir. After having won his

Doctorate of Engineering with high honours at Sydney University, he was appointed Chief Engineer for the New South Wales Government. In 1911 the question of extending both the main and the suburban railways on each side of the harbour was the subject of a long and carefully debated inquiry, which at one stage strongly supported the idea of a tunnel in preference to a bridge. In Dr Bradfield's mind a tunnel was fraught with too many difficulties. The high land of the northern shore being only one of his objections. An advocate of fresh air and sunlight where possible, he also argued that a tunnel posed problems of ventilation and drainage as well as a high fire and accident risk. He strongly advocated for a bridge.

In 1915 a final report to this effect led to the passing of an Act authorizing the extension and co-ordination of both the city-underground and Northshore railway systems together with the building of a high-level rail and road bridge across Sydney Harbour between Dawes Point at the entrance to Circular Quay to Milson's Point on the Northern shore.

The outbreak of war in 1914, delayed tenders for the construction of a bridge and the government sent Dr Bradfield overseas to inspect Underground rail systems and bridges. On his return he prepared specifications for a bridge to span the harbour. Of 20 proposals from six different companies, the contract was awarded in 1924 to the British steel and engineering firm, Dorman Long and Company, of Middlesbrough. Their consultant and chief design-engineer was Ralph Freeman, who had been responsible for the design of two of the most notable bridges in South Africa, including the high arch over the spectacular Victoria Falls gorge of the Zambezi river. Freeman submitted no fewer than seven alternative designs suited for Sydney Harbour, based on Dr Bradfield's report and requirements. They provided a choice between a cantilever or an arch-type bridge. His two-hinged arch design, together with six Approach spans, was accepted. The structure also included granite-faced pylons and piers designed by Sir John Burnett and Partners as well as 52,000 tons of special steel to be fabricated in Sydney to employ both Australian and British workmen. In the currency of the day the contracted price for the Bridge and its six approach spans was 4,217,721 pounds.

As several years were to pass before the long awaited bridge would take visible shape and eventually change the daily travelling habits of the water-side population, the terms and detail of the contract were soon forgotten. All that now mattered was that Sydney was to get a bridge. The man who had conceived, planned and argued successfully for its construction was Dr John Job Crew Bradfield — Australian-born, master-boss, Government supervisor and the Bridge's proud figure-head.

THE MAN WHO PLANNED THE BRIDGE

DR. JOHN JOB CREW BRADFIELD
Chief Government Engineer for New South Wales and
Supervising Engineer for Construction and Design

Born in Queensland, Dr Bradfield, from his early schooldays was to win prizes, medals, honours and scholarships. Later he became Bachelor, Master and finally Doctor of Science and Engineering. He was widely honoured in Australia and abroad. As Chief Design Engineer for the Government of New South Wales, he drew up plans for the City's Underground Railway system and conceived the entire Harbour Bridge scheme for which tenders were invited in 1921. He later visited Canada, U.S.A. and Europe, seeking further relevant technical, industrial and financial information.

His long and brilliant career as an engineer is not only remembered for his part in advancing and supervising the building of the Harbour Bridge but also for the design of dams, inland irrigation schemes, and for the less publicised but successful manufacture of aluminium from Queensland bauxite. The Story Bridge, which spans the Brisbane River, was his last major engineering work.

Retiring in 1933 his later years were spent promoting further irrigation projects and tending his much-loved garden in the Northshore suburb of Gordon, where he died in 1944.

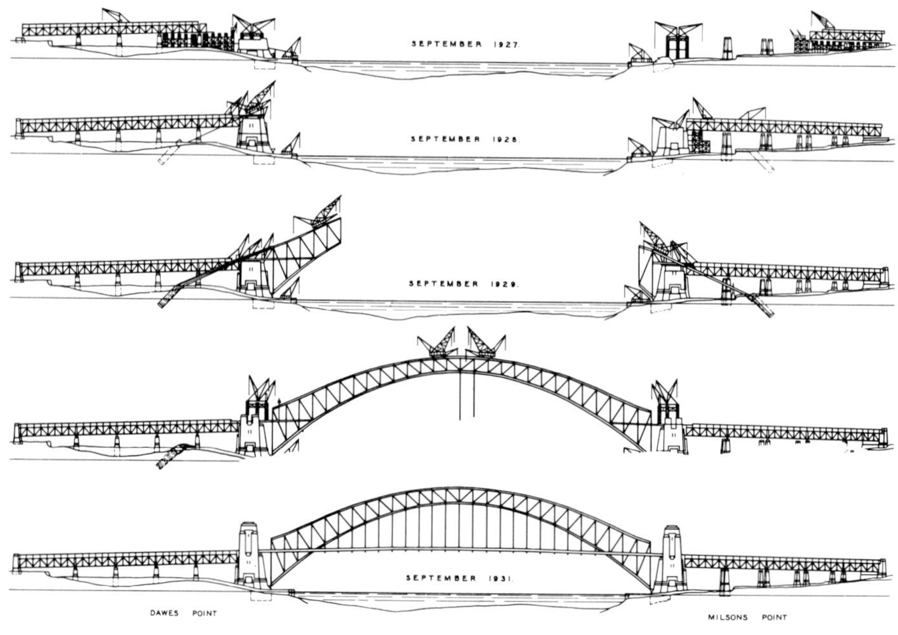

THE MAN WHO DESIGNED THE BRIDGE

RALPH FREEMAN
Engineer and Design-Consultant
for the builders Dorman, Long & Co. Ltd.

Specially appointed to take full charge of the design of the Sydney Harbour Bridge, Ralph Freeman was a senior partner in the British Engineering firm of Sir Douglas Fox and Partners. He was consultant engineer and co-designer of the Victoria Falls bridge over the gorge of the Zambesi river in Africa, and wholly responsible for the design of other bridges on that continent. He also designed the Lambeth Bridge over the Thames and the widely-publicised Newcastle-on-Tyne bridge, which some regarded a miniature forerunner of Sydney's Harbour Bridge.

Ralph Freeman submitted no fewer than seven possible designs for Sydney. His plan for a two-hinged arch of 503 metres was eventually chosen by the State Government.

For his many notable works as one of England's outstanding Bridge designers and engineers, he received a Knighthood. His son, Ralph Freeman, the second, visited Sydney as a young man during the building of the Bridge. He later succeeded his father as a partner in the firm of Freeman and Fox and he too received a Knighthood for his work prior to retirement.

THE MAN WHO BUILT THE BRIDGE

LAWRENCE ENNIS, O.B.E.
Chief Engineer in charge of Construction
for Dorman, Long & Co. Ltd., Middlesbrough, England

Building the Sydney Harbour Bridge was undoubtedly the most challenging contract ever undertaken by this universally regarded company of steel manufacturers. Many engineers throughout the world said that an arch of such colossal weight was likely to topple into the water if unsupported from below during its construction. Such was the burden of responsibility given to its chief engineer. But in Lawrence Ennis, the company had chosen wisely, for he was not a man to be baulked by difficult problems. Besides, he could rely on a splendid team of design engineers, foremen and workers — British and Australian — and never at any stage was the job in jeopardy let alone in danger of collapsing. There were some emotional strains, to be sure, but these were eased by an inherent faith, sound engineering practice and a ready sense of humour. For his many previous achievements Lawrence Ennis had been awarded the Order of the British Empire.

A 'Pommy' new-chum gets a racing tip from his 'Aussie' mate at the Bridge workshops.

On land once granted to James Milson — a free immigrant from Lincolnshire — the approaches to a great bridge are being built. And here today by Milson's Point where scores of old houses have been pulled down, is a sweep of open space, and high above it the Gallipoli Legion Bowling Club.

January, 1928.

'You won't see much yet,' said the Customs Officer on the Darling Harbour wharf as — chalk in hand — he checked and marked our baggage. He had been telling us, with no little pride, that Sydney was building a bridge. 'Biggest arch in the world,' he said. He seemed as fired by the idea as, inwardly, I was. I visualized the splendid subjects such a project would inspire. Years of work in America had nourished my admiration for the art and philosophy of Joseph Pennell and the way he recorded the industrial scene; including the steel mills, bridges, mines, and back in 1914, the building of the Panama Canal. All formed part of his contribution to the Wonder of Work — Today's Work. These subjects were still vivid in my mind and I knew at once that I must somehow get involved with the building of this new Sydney Bridge.

The Customs man had suggested I go down to Circular Quay to see the activity around Dawes Point, thoughtfully adding that if I was interested in a job on the bridge the best man to see first was Dr Bradfield. 'Engineer in charge for the Government,' he said. 'Lives on the Northshore like me. Often on the same train going to Milson's Point and catches the same ferry as me for the city.'

Grateful for the advice, our hopes were nevertheless dashed at first when we could see no sign of the bridge. Where was the thrusting arch?

But the story soon unfolded. Much of the old Rocks district around the Quay and the hill above had been ripped out by its roots. Gangs of men, with monstrous steam-shovels, cement mixers and drillers were still chewing out a route for the approaches to a future bridge. There were muffled explosions of rock-blasting. A cluster of busy cranes towered around Dawes Point where the first bridge-pylon was under construction. Across the harbour at Milson's Point it was much the same, with a long line of corrugated-iron workshops, a shipping dock, cranes and a freighter discharging a cargo of steel. The flurry of activity was so intense on both sides of the harbour that I determined to put my case to Dr Bradfield as soon as possible.

Some months, however, were to elapse while I pursued what might be claimed a legitimate share of Sydney's available free-lance art market. We settled in a Watson's Bay house, complete with studio, and an uninterrupted view of the Harbour's Northern headland. The Bridge site was completely out of range, yet such a house with a tram stop nearby and a ferry service at the bottom of the hill made an ideal base from which to hammer out a new life in a city I had known previously only on a few short visits.

What a year 1928 was for Australians. By mid-February everyone was talking about the wonder-boy from Bundaberg, an unknown Queenslander, who, un-announced, had landed in Darwin having flown a small plane all the way from London in only ten days. Here was Australia's lone eagle, another Charles Lindberg. Wherever he went he was met with a hero's welcome. The nation was caught up in the excitement.

In May, that year, I knocked on the door to Dr Bradfield's elegant office in the fine old sandstone building on the corner of Bridge and Macquarie Street which

Sept. 1930
R.E.C.
The Hydraulic Riveter.
Sydney Bridge Workshop.

The most vital organ in the anatomy of the bridge is its bearing-point (opposite). There are four of them, two on each side of the harbour. They are mounted on deeply-sunk rock foundations. Each consists of a hinge-pin mounted on a massive steel base, which distribute the load of the arch (a weight of some 80,064 tonnes). The pylons against which these bearings are mounted were not built to support the bridge as some people have thought. Despite the statistics, these pins which support the heels of the bridge made a fine design, and a moment of truth for Dr Bradfield.

housed the Chief Secretary and the Department of Public Works. The Doctor was not the physical giant my imagination had envisioned. He was a slight figure of medium height wearing a neat suit with an old-fashioned wing-collar and conservative tie. Seated as he was behind an enormous desk it seemed to accentuate his strikingly large head and his sizeable moustache. He was gracious enough to glance at samples of work I had brought in the hope that he would see the value of obtaining an artist's impression of the Bridge — *his* bridge — now under construction. There was a long pause. In his quietly spoken manner he remarked that strangers were not welcome on the site. 'But wait a minute,' he added and rang a bell. There was a knock on the door, a man entered. After a while I was introduced to one of his senior engineers then, passing me back my drawings, I realised the interview was over and I was escorted to another room. To my surprise, arrangements were made for me to meet the contractors. The Doctor hadn't been so difficult after all. He'd given me the challenge I'd sought and I was grateful.

Working inside the builder's iron-walled workshops at Milson's Point in the rumble and roar of machines made any kind of conversation beyond shouts and gestures impossible but when the engineers and foremen saw what I was doing, I was soon left to take care of myself. Regulations were casual, though I was sometimes warned, albeit aimiably, to 'shift my bloody carcass'. Workers on the site wore only a modicum of safety gear, such as special belts for riggers. Occasionally you'd see a man with gloves. In the shops or later on the arch itself, they invariably wore what they might have worn in their own backyard, the boots, hats, singlets, sweaters and coats they possessed, and they took their chances. Protective helmets were unheard of.

For an artist working around the site there was plenty of subject . The 'heavy' shop, some 65 metres long, was dominated by a giant steel cutter with an awesome bite. Here too were the pneumatic riveters, the screaming drills, the shapers, planers, pounders and hammerers. When all were going full power, the long iron building would shake and roar like a wild thunderstorm. Clusters of men suddenly caught in momentary flashes of light could be seen wrestling with mighty bars of steel, feeding them into machines to be cut, shaped, drilled or riveted. Later, they would be stacked away, each with its special lug for lifting it into its eventual position on the bridge. In the second or 'light' shop, there was a second overhead travelling crane

This drawing shows how the cables were
attached to the arch and, like reins, held it
from plunging into the water. This is the
North Shore abutment.

and a small floating-dock holding a flat-bottomed barge on which all finished parts
were loaded before being ferried to the bridge site.

Only at knock-off time was this bedlam subdued, though never completely. The
topic of conversation, if any, would certainly not be the bridge, but more likely
the odds on next Saturday's favourite starter at Randwick. But that changed in
June when news was flashed — first from Hawaii and then from Fiji — that the
daring attempt to fly the Pacific from east to west by Australians Kingsford Smith

and Charles Ulm with a game crew of Americans, Harry Lyons and Jim Warner, was on. They battled their way, often flying blind, in an old, rebuilt, three-engined crate they'd proudly named the Southern Cross to eventually reach Brisbane and create another 'first' for Australia! What heady times, with more records soon to follow and by women pilots as well. Each and all were widely discussed at work, at home, on the ferries and the trams. As 1928 drew to a close, the bridge builders, now almost a forgotten tribe had reached the water-line and were ready to raise the first end-posts of the great steel arch.

Though the two pylons which faced each other across the water were still unfinished they were already 60 metres high, dwarfing every city building. Some critics dubbed them an excrescence, being unaware of their importance. They were not there to support the bridge or hold it up as so many thought. That would be done by four sets of steel bearings set firmly at the base of the pylons. But they did have a useful function because, at their present bridge-deck level, they provided a launching stage on which was raised the two great creeper-cranes on their respective travelling platforms. These were the 'mother-cranes' and as much a symbol of the bridge during the building as the pylons themselves, their outstretched arms pinpointing the progress of the arch like the feelers of two busy and purposeful spiders threading together a gigantic web.

'See them 'inge-pins?' shouted a foreman with a Yorkshire accent, 'Bridge — she won't stand 'oop widout 'inge-pins!' This of course meant I must not neglect to draw them. The hinge-pins he had pointed out were one of four sets of steel bearings anchored at the base of the pylons. Each reached high above one's head and they would ultimately carry the weight of the entire bridge, some 28,000 tonnes, a staggering thought for the points of contact seemed absurdly small. In the language of the surveyors, their positioning was the one true and fixed reference when aligning the arch and upon their pinpoint accuracy would depend the success of the ultimate joining of the two halves. The Yorkshire foreman was right: no record of the bridge would be complete without a drawing of those 'inge-pins'!

'See the cables, *that's* the way they hold up the arch!'

The builder's contract specified that nothing should block the free passage of ships up or down the harbour during construction, so the two halves of the arch had to be held in suspension until they met in the centre. How could they resist the law of gravity to prevent thousands of tonnes of steel from crashing into the water? The hard sandstone of the harbour bed was the answer. Into this bed of rock, some distance back from each pylon, tunnels were dug, each shaped like a wide horse-shoe and lined with corrugated sheeting against which steel cables could be passed and their ends attached to saddles mounted on the head of the first posts of the arch. Never had such a huge and heavy arch been supported in this way. All would depend on the holding power of those steel cables, the risk being that any unequal distribution of the load would increase the strain on a particular cable, causing it to snap, so increasing the load on another cable which in turn would give way and transmit its load to break still another until finally the whole structure crowded with workmen would collapse. A somewhat hard-headed engineer, when giving me this horrendous picture one day after the two ends of the arch had been joined, added with a real sigh of relief: '. . . and thank God those bloody cables held!'

How well I remember moments, when innocently sketching aloft, hearing the distinct thrumming like a cello string being stroked as more and more cables were attached to hold the growing weight. Not to worry, I was told, it only meant that a particular cable was responding as it should to the strain imposed on it by the creeper-crane moving across the forward tip of the arch.

We were now well into 1930. The two pylons, being only then up to bridge-deck level over which the cables were passed into their horse-shoe tunnels, made a good picture. Commenting on this, one of the engineers suggested there should be subjects worth recording from the top of the arch. He had his camera; I had my sketching gear. Climbing to pylon deck level he was in no mood to dawdle and crossing a narrow plank we reached the first end-post of the arch. Attached to its side was a slim, vertical iron ladder — the kind of ladder that inspires safety in some and holy terror in others. I must belong to the latter for burdened as I was with sketching gear and with only one free arm, it was some time before I joined my companion on the top chord. Standing on that platform of steel, little more than three metres wide, seemingly ending somewhere in the sky! Grand, so long as one watched one's step, for the incline was steep. There were a few broken-down duck-boards but no hand rails. Parallel with the top chord of the arch was the lower chord and forming an enormous arc made up of what the engineer called a series of box-girders connected with vertical struts all diminishing in height as we plodded skywards. Seen in profile every box-girder looked like an enormous capital N. On one of the lower chords a group of men were heating rivets in an oil furnace and I was eager to make a sketch but my guide insisted he had more to show me. On we clambered until halted by the rear-end of the great creeper-crane poised among a tangle of attendant cranes and derricks. A fine lot of subjects here, but just as my friend clicked his camera, the noon knock-off siren screamed to be followed by a scramble of men rushing to board a wire cage which was promptly lowered to water level where a barge was waiting. Returning to the spot where I had hoped to make a study of the riveters I told my guide I'd wait till the men returned. Pointing to a large sack of rivets nearby he suggested I use them as a wind-break and settle in comfort. For the

A view of the advancing arch, as seen from
Observatory Hill.

In a situation like this it is an
advantage to have a friend
along side with a camera. My
drawing is a reconstruction,
partly from memory, as I look
down from the top.

next half hour I had the bridge to myself or so it seemed. All was mysteriously quiet. Just a plank of cold steel pointing skywards, a bag of rivets and the harbour far below — an experience never to be forgotten as I watched tiny ferries spinning their wakes through the broken pattern of an unfinished bridge, and the city stretching its long arms out across the hills to the ocean on one side and the Blue Mountains on the other. At last the men returned to their posts and I got to work. An hour or so later, my sketches made, I slowly picked my way down the sloping back of the arch. Every Sydney rooftop felt like a magnet trying to drag me to my doom. 'Nothin' to it! You'll soon be racin' up and down,' said one of the riggers. If I envied them as they scuttled down the back of that arch, I chose the relative security of being hoisted or lowered in a swaying, crowded cage.

The gradual weaving together of those thrusting arms of steel, the hoisting aloft of every part of the bridge from little flat-bottomed barges at water-level, was the delight of the photographer; men like Henri Mallard who caught the sheer physical drama of the construction and even made a motion picture of it alone and under the most difficult conditions. There was Cazneaux, who never tired of the subject, waiting patiently for a telling moment. And there was the Rev. Frank Cash, whose unique contribution was a day-by-day study from his rectory at Lavender Bay of the great arch as it flung its arms ever higher across the water.

No guards or hand-rails! Just a 3 metre wide steel plank sweeping grandly towards the clouds.

Making ready for the lift of a steel beam by
the creeper-crane. In the background are the
Bridge Workshops by the curving line of the
Northern Approaches.

As the two arms of the arch come closer together, workmen lay a track
to support the creeper-crane as it moves forward along the top chord.
By now, the men are used to working in all weathers on a steel platform
three metres wide and more than 100 metres above the water and not a
single guard rail. I used a camera to record the good figure-grouping.

A few artists tackled isolated phases. One recalls the splendid oil of the Southern Approaches by Percy Lindsay; a series of etchings by Sydney Ure Smith as the bridge was nearing completion, and the bold works of John D. Moore about the same time. The Bridge exerted a particular appeal to several women painters, who translated it in colourful, glowing abstractions. Works by Grace Cossington Smith, Dorrit Black, by Jessie Traill and Vida Lahey, are particularly remembered. Most artists, however, waited until the best of the drama was over and the bridge was completed. The workshops, the details of construction and the immensity of the effort were overlooked. Yet, to me these subjects were as exciting as the bridge itself

One recalls how Sydneysiders particularly the ferry-goers, would speculate about the joining up of the arch. It became a daily topic, as important as the weather. Bets would be taken, for somehow the word had got around that the two halves of the arch were out of alignment and would never meet. It was Lawrence Ennis, the big man in charge of the construction who, in his dry and humourous way, met the speculation head-on. When addressing a public meeting about that time he casually referred to some rumours he'd heard that the arches were out of alignment. His firm, he said, had come to Australia to build a bridge across Sydney harbour. They had started from both sides, he said, and were reasonably sure it would meet in the middle. However, if it didn't, then they'd just go on building until they reached the other side and Sydney would get *two* bridges instead of *one*.

Bets were lost and the rumours subsided. The first meeting of the lower chords of the arch took place as planned and almost unnoticed on the afternoon of August 7th, 1930. Luckily I got a drawing of the event working from a lighter on the southern side of the harbour.

That same day, a news reporter was up on the arch and watched Lawrence Ennis step across the gap from one side of the harbour to the other. His chief erection superintendent Hipwell followed and they shook hands looking out across the water to the city and the ocean beyond and they knew the first vital part of the job had been done, though there would be no relaxing until the two halves were securely bolted together. Seen from the water below everything appeared in order. A day or so and linking pins could be rammed home and the arch would be firmly joined. However, due to a marked difference in expansion of the metal, heat from the sun affecting the northern side of the arch more than the southern, closing the gap would depend on equalizing temperatures of the steel throughout the structure. It had been planned that at the point of meeting the two ends would be about a metre apart and that slackening some of the cables would reduce this to within a tolerance of 115mm, calculated by the engineers. It was the uncertain winter month of August and already days had been spent making adjustments when on the night of August 13, a fierce gale struck Sydney. Men working aloft claimed that the two arms swayed visibly. At one stage the gap might close then widen again as the temperature changed. For days and nights workmen battled aloft with hydraulic-jacks, as telephones rang and temperatures were checked from points all over the structure in the hope of ensuring a perfect alignment. For much of that time,

39

other and less familiar figures would be seen moving among the workmen — checking, inspecting, sometimes voicing opinions. There was the solid, unmistakable Ennis, on whose reputation the success of the project depended. Hipwell, his superintendent engineer, and standing to one side the ever alert, indefatigable Dr Bradfield, whose early dreams of bridging the harbour were at last nearing realisation. More aloof, perhaps, but showing equal concern for every detail was Ralph Freeman, the bridge's designer, who had arrived in Sydney to be present at this crucial period.

Ingenuity, muscle, patience and guts finally won the battle. Nearing midnight on August 19, the two arms were locked together, steel bolts rammed firmly into their sockets and the strain was over. The arch no longer depended on the cables for support. The bearings at the base of the pylons would now carry its weight for all time. For every man on the bridge that night, the strain had been critical. There was

40

Riggers, like the proverbial human flies, guide
a mighty steel beam aloft as the Arch nears
completion. Beneath them — way beneath
them — the busy ferries ply to Circular Quay.
The lines from the work boats below — heavy
as they are — are so long that they are blown
like cotton in the wind.

a pause before Ennis spoke: 'Well boys,' he said, 'that's that and thank God she's
home!' High above a sleeping city the bridge builders shook hands. The job was done!

The Union Jack and the Australian flag were hoisted from the jib of the creeper-
cranes and as the sun rose over the harbour on the morning of the 20th of August,
1930, the victory of the arch was heralded far and wide by the sirens of ships and
the tooting of ferries as cheering crowds went on their way to work.

The restraining cables, those bands of steel which for 20 months had supported
the arch were now slackened off, and falling one-by-one from their anchorage, slid
across the back of the pylons in a tangled coil, some still twitching and writhing as
if their life was not yet over. Indeed, though most were shipped back to England,
others were re-employed and now support the William Taylor suspension road-
bridge across the Brisbane River at Indooroopilly.

In April, 1930, as the two halves grew closer —
rumours abounded that the arch appeared out
of alignment. It was in fact merely an optical
illusion, because when they were ultimately
joined following a critical period, due to the
expansion and contraction of the steel — the
two arms were firmly secured and the full
weight of this colossal arch could then be
carried on its four bearing points at the base
of the pylons.

The arch secured, the two creeper-cranes having met face to face at the centre, now worked slowly backwards down the arch lifting the long, thin deck-hangers one by one that would carry the deck. Perched up there among the clouds, were the drivers of those spider-cranes some kind of supermen, I wondered? What cunning did they employ as they tucked each long deck-hanger so neatly beneath the lip of the arch? One day I met one of them. To his mates he was just Bill, and he too liked to draw a bit. He showed me a number of well drawn ships, all handsomely rigged. 'A hobby,' he said, 'I've sailed around a bit, and when I'm up there in the clouds, like, I get thinking about ships. Keeps me from getting bored.' And how did he manage to tuck those skinny deck-hangers under the lower chord of the arch with his crane? 'Simple,' he said. 'We don't bother about the hangers. We lift the cradle to which the hanger is attached and the riggers do the rest. They fit and rivet them under the part of the lower chord we can't reach, then we send the cradle back and pick up another. They tell me,' he added, 'that one of the engineers worked it all out on his son's Meccano set. Well, fair enough, because it works!'

So, day-by-day, week after week, and for the next nine months, Bill and the two creeper-cranes backed down the arch hoisting first the hangers to which were attached the steel bed of a future concrete and asphalt roadway six lanes wide, plus four standard-gauge rail-tracks (two for electric trains, and two for trams) and two outside walkways. In all the widest single-span traffic bridge in the world! The cranes were eventually dismantled. I would like to have known what happened to Bill and his drawings.

The pylons, which had supported those cranes were now extended to their full height in the form of twin towers, each pierced for a railtrack and a footway. The granite which surfaced them was hewn from a quarry near Moruya on the South coast, mostly by men recruited from Aberdeen, Scotland's granite city. A special township was built to accomodate them and their families, together with a school, a store, and post office. But why granite? To quote the words of Dr Bradfield: 'Future generations will judge us by our works; for that reason despite the additional cost, I have recommended granite. Concrete is man's artificial stone but granite is the strongest and most enduring and will withstand the ravages of time for thousands of years.'

By the end of 1930, Australia was in the full grip of the Depression. It had started late in 1929 with the collapse of the New York Stock and share market. Australia which relied heavily on the sale of her primary products seemed at first in a safer position. But she also depended on heavy overseas borrowing. When the financial wells dried up, what had started as a trade recession the previous year now struck the country with the full cruel force of a depression. Tens of thousands were thrown out of work, crushing the weak hope that conditions would soon return to

Robert Emerson Curtis - April 1930.

normal. They were traumatic times for most, yet through it all the work on the bridge never stopped.

In 1930 I made my first lithograph, a drawing on a zinc plate, showing the first stage in the joining of the arch. I made another of the bridge-deck nearing completion and both subjects were duly printed. By now most of the free-lance artwork which I depended for a living had dried up and there was no money to meet the $4 weekly rent of our Watson's Bay house. Buying a tent and some essential camp gear we left Sydney and the Bridge and spent the next six months across the border on the high slopes of Queensland's Granite Belt, selling an occasional drawing and doing some fruit-picking, our existence was as close to earthly paradise as I could conceive. It was an illusion, of course, for such a state can never last long. With the thought of the Bridge still nagging my mind I returned to Sydney at the invitation of a friend together with the promise of co-operation by a printer to let me use his equipment to produce a further series of lithographs. No-one in their right senses would aim to publish a book in the depth of a Depression, but together with my friend the printer, we took a chance and issued a folio of lithographs for the price of a guinea ($2.10) and Dr Bradfield kindly penned a Foreword. But who had money to spare to buy them? Some retail bookshops took a chance. There were a few doctors with an engineering bent. And some wanted a record of the bridge on which they had worked. My wife now joined me in an all-out sales campaign and together with a camping mate and his old car, we set off in high spirits for Melbourne, only to find that the building of Sydney's great bridge excited them little and stirred their pocket books less. However, not without hope and plenty of time to spare, we took to the road, walked back to Sydney, meeting many friendly and courageous people along the way. We also saw the granite quarries of Moruya and talked with the brawny lads and their wives from Aberdeen. We arrived home after a month to find the Bridge all but complete, together with its smart new stations on the northern bank, and its carefully groomed road approaches, verges, steps and plantings, all reflecting the Bradfield genius for detail. To prove what substantial weights the Bridge could carry, its rail tracks were lined with some 69 ancient, long-retired steam locomotives parked buffer to buffer between the two pylons, a sight to tickle the hearts of engine lovers, young and old!

Sydney Harbour Bridge
August 6th. 1930.

Honour to the men doing their daily shift with a drill or a pick who curse the dust and the heat; caring nothing for the homes they are knocking into rubble. And in the workshops where giant slabs of steel are cut, planed, drilled and pounded — it's all in a day's work. Cement mixers and the rock masons high aloft setting granite slabs on a pylon's naked wall. The men who dig the tunnels, lay the cables, climb the arch, drill the holes, heat the rivets and force them 'home'. Honour to these men who are building a Bridge. These men are building a bridge. They shout and curse and laugh; pounding steel; heating rivets; forcing them home with muscle power and sweat. High aloft in their cabins, the crane-drivers look down to the men below swinging hammers; riding the beams; but they too have their problems. Every man gives something of himself to this bridge of steel and fire, built with the toil, the sweat and the guts of a thousand men.

The story explains how these immense deck-
hangers, shown on the cradle, are lifted and
attached beneath the lower chords of the arch.

On a foggy morning high above Pitt Street the
two 'mother-cranes' have woven a 'Gothic Arch'
in the sky while in the street below life goes on.

About this time, Kingsford Smith and Charles Ulm, now world famous pioneer aviators, were making every penny they could in pursuit of their dream to establish a national Australian airline. They planned to fly to England and purchase suitable aircraft and to hasten their plan 'Smithy' was offering joy-rides over Sydney for $1 a head. The temptation was irresistable, and Smithy gladly made several runs for me over the Bridge, so I might visualize it all from the air on the forthcoming day of the opening celebrations.

Building the Bridge, and everything connected with the work, had taken eight years. 'Does that mean the Bridge will last forever?' I asked Lawrence Ennis.

Cheering crowds line the decks of ships and
ferries on the morning of August 20, after the
dramatic night of the joining.

'Putting it bluntly, the Bridge will last as long as they keep the rust out of it, short
of an earthquake or a bomb,' he replied. It sounded simple enough at the time,
but it wasn't.

It has now been exposed for over 50 years in every kind of weather; salt-laden
air, beating sun, driving rain, hail, wind and smog. To date it has been repainted
five times, each repaint taking an average of ten years. First, a careful check for
rust; then a scrapedown; then a red lead primer; an undercoat; and a finishing
coat. A staff of painters, riggers and iron-workers are permanently on the job.
There are also more than 100 sliding bearings at the expansion joints to be main-
tained. Would the Bridge look any smarter in red or gold; silver or blue? For half
a century Sydney has accepted its sober, business-like grey. It's easy on the eyes,
and it blends with the environment.

One of my first lithographs; 'The hanging of
the deck'. What appeared to be a unique per-
formance of the creeper-cranes until they ex-
plained to me that they don't lift the hangers —
we only lift the cradle to which the hanger is
attached — the riggers then rivet them into
place under the lower chord of the arch.

"Building the Bridge"
Sydney. Dec 1930.
R.C.C.

Lining the rail-tracks, buffer to buffer, 80 retired
locomotives test the strength of the bridge-deck —
the delight of engine lovers — young and old!

During the building years, at least eight men are known to have fallen to their
death from the Bridge. There were few safety controls; few protective rails aloft
to clutch in sudden wind gusts, storms, rain or fog. One rigger, missing his foothold, managed to hit the water with his body hunched like a cannon ball. He
eventually rose to the surface, was picked up and survived with only shattered bones.

For many generations, the 'Gap', or 'going over the Gap' was a time-worn music
hall joke, an 'Aussie' joke with a macabre twist. Hundreds of sight-seers would
climb the hill to the 'Gap' high above Watson's Bay, to look far below at the sea
pounding the jagged rocks, roaring like thunder. Many a disturbed soul has been

The last of old Argyle Cut. Not until the Bridge was
nearly built was this historic Cut widened. In the
original rock were anchorage points that once carried
iron rings to which convict labourers digging the
tunnel had been chained.

The last of old Argyle Cut.

tempted to jump. Now the Bridge was open, a few desperate souls took a new plunge that seemed an easy way out, until the high restraining wire guard was erected.

If, some fifty years ago, the folk of Bayonne, New Jersey, USA, could claim they had built a single-steel arch a few centimetres longer, or that the world has since built greater or more unusual bridges, does it really matter? If our grand old Steel Lady is now a trifle out-dated when compared with the light-weight, clean cut spans that a new generation of engineers have devised, she has nevertheless served us day and night for half a century, and faithfully.

When the distinguished American, Miss Helen Keller, visited Australia some years ago, she asked the Governor-General the length of the Harbour Bridge. He could not oblige, nor could many others. When the writer, Shawn O'Leary heard of this, he promptly questioned a number of his friends. Not one of them had the correct answer. To quote his own words: 'I turned to the experts.' At the back of this book, you too will find most of the answers to your technical queries, compiled by the experts of the Department of Main Roads, which maintains the Bridge, it's tolls and controls its road traffic.

Idle and dispirited ships, anchored for months in Rose Bay during the Depression waiting for cargos that never turned up.

Finishing off the Bradfield Highway (opposite) some 3 months before the Bridge was opened to traffic.

The lithograph on pages 56 and 57 was made some months
before the Bridge was officially opened, before trains, trams,
motor cars and the crowds made their first daily pilgrimage
across it. Until then, Sydney's sixty or seventy faithful coal-
burning ferries held sway while anyone bound for the city
from Milson's Point had to race down the escalator and face
the morning stampede at the end of the railway line before
boarding the boat at Circular Quay.

'Without pylons' was an alternative proposal by
Ralph Freeman. The pylons, which are now
part of his bridge, were designed by Sir John
Burnett, in consultation with Ralph Freeman
and Dr Bradfield.

The cutting of the ceremonial ribbon was a
play in two Acts. It had all the elements of
a comic opera — more politely referred to as
a 'contretemps' — and which failed to mar the
proceedings, and in fact, added great excitement
to them. Captain Francis de Groot — an officer
and member of the New Guard — will long be
remembered for his spirited dash, mounted on
a horse and with sword drawn, slashing the
ceremonial ribbon and scooping the honour
from the redoubtable Jack Lang — arch enemy
and political target of the New Guard.

GAME
ENNIS
LANG
BRADFIELD

The Opening

In all its colourful waterside history, Sydney never witnessed a spectacle to match the opening of the Harbour Bridge. It was a day of pomp and dignity in brilliant sunshine with a dash of comic opera for good measure. By now Australians far and wide knew of the longest, widest, single-steel arch in the world, and the first to span Sydney Harbour.

With the associated underground rail and road approaches, the entire scheme had taken eight years to build. Yet despite the economic trauma of the times, neither strikes nor the Depression itself had delayed the work. Nothing was now going to stop the people flocking in thousands to enjoy the celebrations.

His Majesty King George V was represented by the first Australian-born Governor General, Sir Isaac Isaacs, a Victorian Prime Minister Joe Lyons was followed by Sir Phillip Game, Governor of New South Wales, whose Parliament was led by the Premier, Jack Lang. Lang was a colourful giant of Australian Labour politics, whose name alone could arouse passions of devotion among his supporters and the bitterest hatred in those who were convinced he was ruining the country. Jack Lang was to officially open the Bridge in a ribbon-cutting ceremony, an honour not to be stomached by the New Guard, a semi-militant right-wing band bent on toppling him. They connived to thwart him in the act.

Moments before the opening ceremony was to take place, a stalwart New Guardsman, Captain Francis de Groot, decked out in officer's uniform, spurred and mounted on a borrowed race-horse, slipped from the rank of the Governor General's Horse-guard. With sword drawn, he charged the official barrier, and slashed at the ribbon. "In the name of the decent citizens of New South Wales, I declare the BRIDGE OPEN!" he shouted before being jerked smartly from his horse and tumbled to the ground. He was led away and later placed under strong guard at a reception house for the mentally unhinged.

[Many days later, there being no evidence of insanity, he was released, charged with malicious damage to property, and fined the sum of ten pounds.]

The redoubtable Jack Lang eventually performed the official opening ceremony, with re-joined ribbons, watched nervously by Sir Phillip Game, Dr Bradfield, Lawrence Ennis; the Lord Mayor of Sydney, Alderman Walder, and others. After a 21 gun salute the bands struck up as the crowds milled around or joined the long procession of floats the full length of the Bridge. At this moment several ocean liners, their decks crowded, sirens tooting, steamed through the shadows of the great arch into a sunlit harbour lined with packed ferry-boats and private yachts. High above, a squadron of the Royal Australian Air Force rent the sky to ensure that no-one in the city missed the hulla-ba-loo! A few rode the first electric passenger train across the bridge (for a return fare of ten shillings) while thousands

After the ceremony of cutting the ribbon and a 21-gun salute, crowds stream across the newly opened Bridge as a procession of smart ocean liners is led down the Harbour by the 'Captain Cook' — Sydney's smart and popular pilot-boat of the day.

The lithograph on pages 64 and 65 shows a view from
a Woolloomooloo rooftop. To the right are the docks
and an Orient liner setting off for Great Britain. On
the far slope of the Botanic Gardens — Government
House and the newly opened Bridge, and across the
water a fast expanding North Shore.

Before receiving a Knighthood, Charles Kingsford
Smith charged a modest 10 shillings ($1) to fly me
around the harbour to visualise what the Bridge
would look like on the festive day of the Opening
Celebrations. These were tough 'barn-storming' days
for Australia's Hero of the Air, and his flying partner
Charles Ulm, whose ambitious plans were to start an
Australian National Airline. My drawing shows
Sydney's pilot-boat 'Captain Cook' leading the P&O
liner 'Maloja', followed by the 'Orford' of the Orient
Line under the arch of the Bridge.

Not everyone welcomed
the Bridge. This letter,
with letterhead removed
reflects the anger that
more than a few felt at
the time.

Sydney, June 16th, 1932.

R. C. Curtis, Esq.,
     62 St. John's Avenue,
        GORDON.

Dear Sir,

        Your book of lithographs of the Sydney Bridge was
received at my house.

        I have not even looked at the same, as the Bridge
represents to me the Tragedy of Sydney and reflects in the
highest form the lack of stability of the Sydney people.

        It is 50 years ahead of its time, serves no really
necessary purpose, and the Politicians who put the Bill through
Parliament and the Cabinet who made the contract have placed an
incubus upon, not only this generation, but the one to follow.

        Your lithographs can be picked up at this office.

        Yours faithfully,

        c. a. d. dowey —

more crowded the foreshores after dark to witness the grandest ever of Sydney's fireworks and searchlight displays. On the stroke of midnight the first public motor and cycle cavalcade to cross the Bridge to the northern shore set off. If there was a moment in 1932 when the Depression seemed forgotten it was at the Grand Opening of the Harbour Bridge — made even more memorable by the antics of a man on a borrowed horse.

Throwing a web of steel across the harbour with not a single support from the water below may have once seemed impossible but it had been done. Long before the official opening Sydney gift-shops were overflowing with souvenirs of the event, and every one a front-on view of the grand arch which was to enshrine forever the 'coat-hanger' look! Printed, etched, stamped, cut, modelled, carved, stitched and poker-worked on everything from beer mugs to boomerangs. Such is monumental fame! But the Bridge was Sydney's great love-affair, and its story was flashed around the world.

For some the Bridge had been a deplorable waste of money, but for those who lived on the northern side of the harbour and the thousands who would later follow, it hadn't come any too soon. No more those open vehicular punts or the pell-mell scramble for a foothold on a crowded ferry. Such days were over. For hundreds of

Work begins to 'modernise' the Quay and build new ferry wharves after the opening of the Bridge in 1932.

One of the few quiet on site places for a sketch,
beneath the Southern Approaches to the Bridge
at Dawes Point.

Lithograph on zinc plate - no 35/50
Sydney Harbour from the Heights of Vaucluse - 1932.

In 1932, when the Bridge was opened,
there were four times as many people
living on the south side of the harbour as
on the northern shores, yet there was still
vacant land high above Rose Bay and
Vaucluse. This lithograph recalls to mind
the words of the late Marjorie Barnard,
who speaks of the Harbour 'lying in the
lap of the city — a little inland sea with
its 183 miles of coastline, where the
native trees have gone, except on a few
reserved headlands, the silence has gone,
but the colour and the pattern of the
water still dominate and the enchantment
remains . . . tens of thousands of people
now look out over the water — a play-
ground and a busy port and as beautiful
as it ever was for all that the 'iron-shackle'
of a bridge has been layed upon it . . .'

more distant commuters who had discovered their elixir among the distant gum-scented ridges there was now a direct electric train service to the very heart of the city! When automobiles had running-boards and the roads uncrowded (and many unsealed) who then talked of ecology or feared the dragon of environmental impact? From the era of the felt hat, from the 'horse-punt' and the 'Tin-Lizzy'; through depressions and wars; from pounds sterling to decimal currency; and from miles to kilometres; the Bridge has seen the city grow from a little over a million to a population nearing three and a half millions. And from the shores of what once was Sydney Cove, the birthplace of the nation, we have watched the transformation of the city's skyline, the reconstruction of Circular Quay, and the changing pattern of transport.

At the time of the opening of the Bridge, the eminent Leon Gellert, a Sydney commentator, wrote: "The skyscraper is almost upon us and in a few years Sydney's architecture will rival the height of its Bridge, and for those who would begin a

The Bridge and the new ferry wharves have
been built. The Cahill Expressway and the new
Railway station are almost complete. On Bennelong Point — at the eastern entrance to the
Quay — is the old bus and tram shelter, its
exterior, designed like a fort, which had once
replaced a much earlier Fort Macquarie. Today,
the roof sails of the Sydney Opera House rise
over Bennelong Point and towering office
blocks greet the traveller who comes by sea.

Looking across the Bridge to the
city from the northern shore.
Following the introduction of new
building-height laws, office towers
are rising on both sides of the
Sydney Harbour.

Fifty years on! The Doctor who
conceived and planned it all!

For fifty years, and often strained to capacity, the Bridge has never failed to absorb and provide a continuous flow of traffic. During the 70s, more than 50 million road vehicles crossed the Bridge every year. Traffic-counting equipment registers the number of vehicles using each lane at any point of time, making it possible to provide peak-hour traffic with additional lanes as required. The Cahill Expressway, linking the Bridge with Macquarie Street and the Eastern parts of the city, was opened in 1958. The Western Distributor extends from the Bridge to Ultimo and South and Western Freeways. On the North side of the harbour, the Warringah Freeway will eventually provide access to all suburbs along the Warringah Peninsula. 'The Bridge' say the controllers, 'will never reach saturation point, for it is only part of a roadway. The only thing that will cause the bridge to be useless is the inability of the streets at either end to absorb approaching and departing traffic.'

quarrel on aesthetic grounds, let it be understood that the engineer will be the architect of tomorrow. Perfect engineering, as a thing of beauty survives by its very simplicity. There is no waste. It achieves its purpose by the shortest route, the finest materials and absolute obedience to natural laws. The perfect work of engineering is a perfect object of art and as such will be recognised by the critics of the future!" Has Sydney's 'love-affair' with her Bridge grown a little thin? Is that magnificent web of grey steel — glimpsed through the windshields of thousands of cars jockeying for position as they edge, roll or trickle in the thick of peak-hour congestion — more likely to raise blood-pressure than uplift the heart? Does it help to be told that peak-hour congestion can be avoided by sensible 'timing' of the traffic flow, or that the Bridge could carry twice its present tally of cars, trucks, buses and trains? To quote the Department of Main Roads, who control the Bridges' traffic, its Public Affairs Officer states: "The Bridge will never reach saturation point for it is only part of a roadway. The only thing that will cause the Bridge to be useless is the inability of the streets at either end to absorb approaching and departing traffic." Short of a second or a third Harbour bridge-crossing, surely that sums it up!

Love-affair or past glory, the great arch still ties the twin cities of south and north together. As she reaches her 50th working year she's still the city's noble, uncomplaining traffic-bearer. 'Steel Lady' or giant 'coat-hanger', call her what you will. But as long as the painters keep the rust out of her stays and joints she's structurally sound enough, so the engineers say, to see us through for many more generations.

To put to rest the long debated question of who designed the Bridge, the honour of conceiving and planning for a high level Bridge in conjunction with a City Underground Railway system is the work of Dr J.J.C. Bradfield, and a great achievement. The design of the Sydney Harbour Bridge and one of seven designs submitted to meet Dr Bradfield's plan and requirements, is the work of the late Sir Ralph Freeman. Honour then to both these men, for the Bridge which has received the Australian National Trust's highest accolade for a building to be preserved for all time.

Home-bound traffic moving across
the Bridge, sketched from the rooftop of a Milson's Point apartment
block, showing the broad curving
highway leading north.

Looking north across the harbour — an X-ray image
of the Warringah Expressway and some of its feeder
lanes. On the southern side of the Harbour — high
above the new railway station and the ferry-docks is
the Cahill Expressway, which avoids the city centre
to connect with Eastern Suburbs. The strong, utilita-
rian beauty of the Bridge and the gleaming and sculp-
tured silhouette of Utzon's Sydney Opera House now
meet the incoming traveller to the Quay around which
the town was born.

The great arch of the Bridge rises high above
every hill-side apartment-block overlooking Car-
eening Cove and the Royal Sydney Yacht Club.

Facts & Figures

1815: A Harbour bridge scheme, including fortifications on Dawes Point was proposed by architect Francis Greenway. However, the scheme lacked details of how it was intended to span the channel nearly 1 kilometre wide and 15-20 metres deep.

1857: The first practical drawing for a bridge was made by Peter Henderson, a Sydney architect, who had served in the workshops of George Stephenson, builder of the first railway locomotive.

1878: A floating bridge was proposed by W.C. Bennett.

1879: Seven span truss bridge was proposed by T.S. Parrott.

1913: The Public Works Department accepted Dr Bradfield's scheme for either an Arch or a Cantilever Bridge, together with plans for an underground electric rail system.

1922: Dr Bradfield's plan passed by Act of Parliament.

1923: From 20 proposals, the tender by Dorman, Long & Co. Ltd. of Middlesbrough England, was accepted. The design was chosen from seven submitted for a 2-hinged arch by the contractor's consultant engineer, Mr Ralph Freeman. Its construction was the responsibility of Mr Lawrence Ennis, chief Director of the firm. The contracted price for the Bridge and six Approach spans was 4,217,721 pounds. The total cost of the work, including all Approaches, the resumption of property and incidentals was 9,578,000 pounds.

Before the main Bridge work was started, some 800 buildings had to be pulled down. Some 333 buildings were demolished in the Rocks area, the most notable being the old Scot's Church, erected in 1826, by Dr. Dunmore Lang. A further 467 properties disappeared on the Northshore.

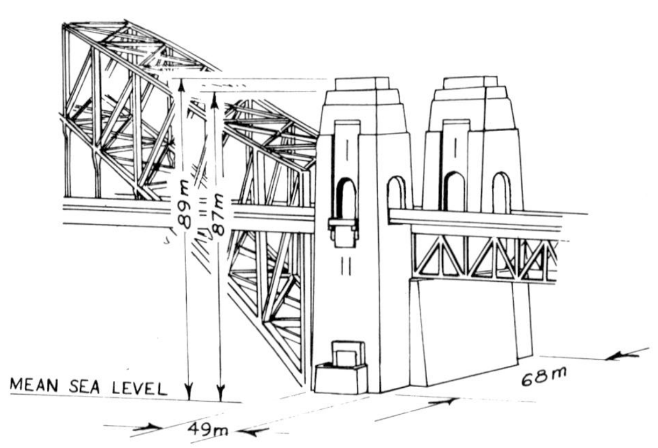

89 m
87 m
MEAN SEA LEVEL
49m
68 m

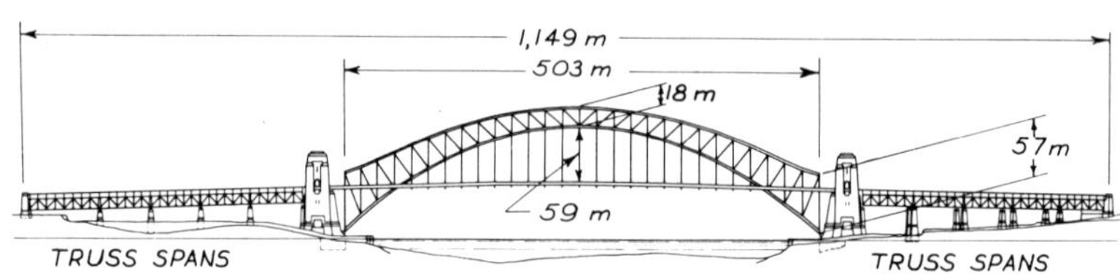

1,149 m
503 m
18 m
57 m
59 m
TRUSS SPANS
TRUSS SPANS

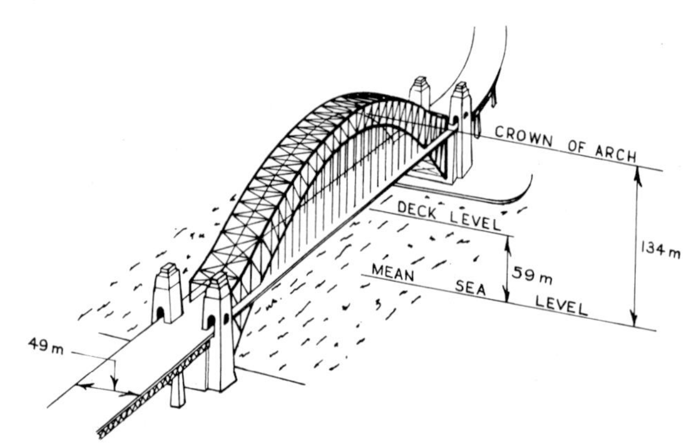

CROWN OF ARCH
DECK LEVEL
MEAN SEA LEVEL
59 m
134 m
49 m

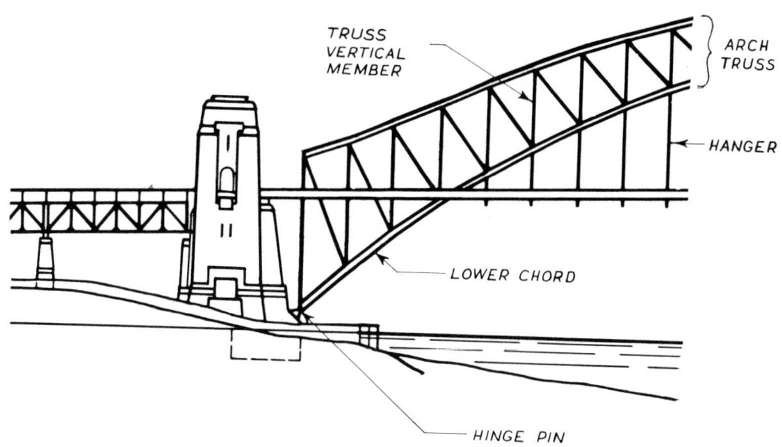

About 53,000 tonnes of steel were used in the fabrication of the Bridge and Approach spans; 10,000 tonnes of which were milled in Australia. Silicon steels in the arch; carbon steel on bridge-deck and approaches. In all as many as 754 workers, mostly Australian, and some from Britain, were employed in the workshops on the Bridge.

The two pylons, each 87 metres tall, were designed by Sir John Burnett & Partners in consultation with Dr Bradfield and Ralph Freeman. They do not support the arch, though they form part of the overall design concept, as well as having various practical uses. Referring to their construction, Dr Bradfield recommended they be faced with granite as the strongest and most enduring of natural building materials, despite an added cost of 240,000 pounds.

To supply the granite blocks used in the construction of the pylons and the Approach piers, a quarry was opened near Moruya on the south coast of New South Wales. A number of highly-skilled quarry-men were brought from the granite town of Aberdeen in Scotland to work the stone, and a small township was built nearby for the workers and their families.

Four main bearings (two at the base of each pylon) take the weight and thrust of the Arch, some 39,006 tonnes. The steel pins on which the Arch rests are each 4.2 metres long, and 368mm in diameter.

During the erection of the Arch, its weight was held in suspension over the water by two sets of 128 steel cables which were attached to the top of the four end posts.

Though the arch itself forms what appears to be a visible curve, it is an optical illusion. There is not a single curved section in its construction.

The combined length of the arch span and steel approach spans is 1,149 metres.

Six million rivets were 'punched home' on the Bridge. All were pre-heated on site in small portable oil furnaces.

The Bridge was opened to traffic on the 19th March, 1932. The average daily traffic flow across the Sydney Harbour Bridge by 1971-72 was 134,700 vehicles, equal to 50.2 million vehicles per annum.

One-way toll collection on Sydney Harbour Bridge came into operation at midnight on Saturday, July 4th, 1970. The toll rates now are:

Vehicles (not otherwise specified) —

    (a) Whose tare weight exceeds 2 tonnes . . 40 cents

    (b) Whose tare weight does not exceed
        2 tonnes . . . . 20 cents

Motor cars . . . . 20 cents

Motor cycle with sidecar or motor tricycle . . . . 10 cents

Solo motor cycle, motor scooter or
horse-drawn vehicle . . . . 5 cents

The new toll charges for south bound traffic were double the old rates, except for cycles, scooters and horsedrawn vehicles which paid 5 cents instead of 2 cents. A flat rate single coin toll for each class of vehicle travelling in either direction was introduced on April 4, 1960. Prior to that date, the toll charged varied according to the type of vehicle and the number of passengers carried.

In conjunction with the introduction of one-way toll collection, toll barriers on the southern approach to Sydney Harbour Bridge were replaced by 6 movable booths and automatic toll-collection equipment was installed at two booths at each end of the Bridge. Toll registration equipment was first used on Sydney Harbour Bridge on January 2nd, 1962, and replaced a system based on the sale of tickets, which had been in force since the Bridge was opened in 1932.

An air navigation beacon was erected on top of the arch in 1949. The beacon is 141 metres above mean sea level. 2 diamond-shaped illuminated navigation signs are provided as a guide for shipping — one at the eastern and one at the western side of the deck at mid-span. There is also an antenna on top of the arch, used for radio telephony communications by the Public Transport Commission of New South Wales.

Grateful acknowledgement is made for the above facts and figures which have been supplied by the Department of Main Roads of N.S.W. who administer and control all operations and services in connection with the Bridge and its maintenance.